Perlorian™ Seasons

Suzanne Green

Doubleday & Company, Inc.
Garden City, New York

Perlorian Cats is a trademark of Satoru Tsuda

Library of Congress Cataloging-in-Publication Data
Green, Suzanne.
    Seasons.
    Summary: Photographs of cat characters engaged in
seasonal activities such as spring fishing and winter
sledding illustrate the variety of the four seasons.
    1. Seasons—Juvenile literature.   [1. Seasons]
I. Title.
QB631.G73   1987   574.5'43   86-16193
ISBN 0-385-23506-2
ISBN 0-385-24007-4 (lib. bdg.)

Note to Grown-ups

The "Perlorian Cats" that you see here are very special animals photographed by a very caring group of photographers led by Satoru Tsuda. The cats are specially chosen for their expressive faces and comfort with the photography sessions.

These photographs are taken at incredibly high shutter speeds to capture a pose and an expression without any discomfort to the cat or cats involved. No artificial substances are used—just love and patience! And the cats seem to respond beautifully to the attention and caring that surround them.

Needless to say, no one should try to do this with cats or kittens on their own. Professional training and proper circumstances should always be involved when working with animals. Your family cat will not welcome treatment it is not used to. Cats are very independent animals!

Spring brings warmer
weather and a chance to go
hiking and camping.

Look what we caught!

Ready to cook our catch.

Bedtime!

Spring is time to start
anything—even a new
business.

Maybe *this* will work!

Oh, well. At least we won't
go thirsty!

Summer is the best time for
an outdoor fair. Let's eat!

Smile for the camera!

Let's ride the Cat Racer.

I hope it's not too fast!

Gulp!

Wasn't that fun?

Summer is the time for surfing.

Off we go.

In summer we celebrate the
Fourth of July.

Fall brings cooler weather
and Halloween.

We're making a
jack-o'-lantern.

How do you like our costumes?

We *love* trick or treating!

Winter brings cold weather
and indoor fun like playing
cards . . .

. . . and playing darts.

It's time to build a snowcat.

Isn't it a fine one?

Winter sledding is lots of
fun. Goodbye for now!